I0436884

The Governor's Suits

The Governor's Suits

A Psychiatric Perspective of Puerto Rico

DR. GUILLERMO GONZALEZ

Copyright © 2007 by Dr. Guillermo Gonzalez.

Library of Congress Control Number:		2006910211
ISBN:	Hardcover	978-1-4257-4426-7
	Softcover	978-1-4257-4425-0

All rights reserved. No part of this book may be reproduced or transmitted in any form or by any means, electronic or mechanical, including photocopying, recording, or by any information storage and retrieval system, without permission in writing from the copyright owner.

This book was printed in the United States of America.

To order additional copies of this book, contact:
Xlibris Corporation
1-888-795-4274
www.Xlibris.com
Orders@Xlibris.com
37899

Contents

Acknowledgement..7

Preface...9

Introduction 11

Chapter One: Colonized Personality Disorder.............................15

Chapter Two: A New Social Class Arises In Puerto Rico
                  From The Colonized Personality:colonized Politicians...........21

Chapter Three: The Colonized Personality.....................................32

Chapter Four: Beyond The Colonized Personality 50

Chapter Five: Predictions...60

ACKNOWLEDGEMENT

M Y DEEPEST GRATITUDE to my companion, Sophia Sotis. The formulation and expression of my ideas became the written word during my interactions with her. Her skills in the Greek, Spanish, and English languages have been a great help to me. Without you, Sophia, it would have taken me another thirty-seven years to write this book.

PREFACE

I CONSIDER THIS book to be a deferred response. I first confronted the problem of describing the Puerto Rican personality thirty-seven years ago. At that time, it seemed to me to be a challenge, and it motivated me to take up multiple studies and exchanges. I felt a shortage of appropriate instruments to help me to answer this question, even after so many years of experience. Describing an individual's personality is not the same as talking about a social group. It is not easy to make use of variables that describe the individual's interactions with his or her physical and social environment. It is even more difficult to talk about how history interacts with the individual, not only the history of Puerto Rico but also the history of humanity. In the absence of a clear consensus in the field of psychology, I have made a very subjective decision to choose and utilize some concepts that are the result of both my personal and professional experiences.

For many these concepts will be original: for others perhaps absurd and incorrect. I have dared to propose that the concept of the unconscious is a reality with an identifiable biological substratum. For me, the individual unconscious, as well as the collective unconscious, reside in our genetic code, better known as our DNA and RNA.

The dimensions I used to evaluate the personality were also selected in a very subjective way. Some of these dimensions are a product of my clinical experience in the practice of psychiatry.

I want to define, prior to delving into the content of this book, the prejudices that have allowed me to answer the age-old question in my mind regarding the description of the Puerto Rican personality. I will be discussing the following variables whenever I describe any social group:

1. Individual awareness of the historical and current socioeconomic reality of the country, as well as contact with reality
2. Emotional maturity of the individual
3. The brain's mode of processing information
4. Philosophical perspective in life
5. Attitude about and contact with the body and physical processes
6. Experiences relative to the coordinates of time and space
7. Attitude in relation to private property
8. Leadership style when solving social problems
9. View of the world
10. Clear understanding of the value, usefulness, and dynamic nature of language

I acknowledge that I could have used more and perhaps better variables to answer such a daring question, but I am not going to delay this process any longer.

INTRODUCTION

I WAS BORN and raised in San Juan, Puerto Rico. I am a physician by profession with a specialty in psychiatry, a profession I have practiced since 1973. I studied in the Puerto Rican public school system. I continued my studies, uninterrupted, at the University of Puerto Rico at the Rio Piedras Campus. I likewise carried out my medical studies and my psychiatric residency at the University of Puerto Rico at the Medical Sciences Campus. I completed part of my studies in the city of New York at State University of New York, Downstate Medical Science Campus at Brooklyn. I moved to the state of Massachusetts in 1992.

This constant movement from Puerto Rico to the United States of America is very typical for many Puerto Ricans. Currently, the number of Puerto Ricans living on the island is similar to the number of Puerto Ricans living outside of it. My family began this migratory process to the United States during the 1930s. These departures from my land of origin have offered me, as well as many others, new perspectives and ideas about Puerto Rico. In this book, I intend to share my ideas and experiences during this process. Much has been written about Puerto Rico from different perspectives.

What distinguishes this from the others is that I will present the ideas and experiences of a psychiatrist during this process. The practice of psychiatry in both countries has given me the opportunity to compare and reconsider the previous concepts on themes related to personality traits found in Puerto Ricans.

It is typical of doctors to study pathological conditions, better known as diseases. We look foremost to defining these states through observation. We then try to identify their causes in order to be able to treat them and cure them, if such a thing is possible. This medical model is what will guide my trajectory throughout this book. I should recognize that it has only been in recent years that psychiatry has adopted this model versus the model that was prevalent before, that of psychoanalysis. It is within these specifications and limitations that I propose to present to you my findings.

In the United States, as well as in Puerto Rico, psychiatrists use the statistical manual of classifications, better known as *DSM-IV (Diagnostic and Statistical Manual of Mental Disorders, Fourth Edition)*, which was created by the American Psychiatric Association. Different mental illnesses, including the so-called personality disorders, are listed in this manual. In this book, we will make frequent references to the latter. In order to be included in this manual, the syndromes have to be associated with and cause dysfunction in interpersonal relationships, as well as in occupational performance.

Throughout my many years of practice, I have carried out thousands of psychiatric interviews and evaluations with Puerto Ricans living in Puerto Rico, and those living in the United States. I have observed, more specifically in regards to personality disorders, the presence of a distinct and unique syndrome in Puerto Ricans as compared to those described in *DSM-IV*. This book deals with that. I have taken the initiative to name it colonized personality disorder. Throughout this book, I will describe its fundamental characteristics, possible causes, and possible alternatives for treatment.

I think that although it is not currently included in *DSM-IV*, if in the future Puerto Rican psychiatrists decide to write their own diagnostic manual, the personality disorder that is described here should be considered for inclusion therein.

This psychiatric syndrome represents a pathological extreme of human behavior, which my professional practice has revealed to me. My point of departure is the many years of practice during which the evaluations of psychiatric disabilities have constituted a significant percentage of my clinical tasks.

I use the *DSM-IV* manual in order to understand these conditions. Nevertheless, what I have discovered in Puerto Ricans is difficult to describe within the categories included in this manual. It is similar to several types of diagnoses, for example, dependent personality disorder and avoidant personality disorder, but it is not exactly the same.

Moreover, it is interesting that these findings show that these personality traits are consistently observed in Puerto Ricans who are not mental health patients. These observations have led me to draw the inference that, in addition to presenting as a psychiatric syndrome in and of itself, these personality traits could also be present in the personality of many Puerto Ricans.

From these observations, first with psychiatric patients and then with nonpatients, I have asked myself if what I have observed in others has some relevance in relation to my own being. This self-analysis has been very revealing, and I have felt that many of these traits have been present in my own personality at different stages of my life, including now. For that reason, I have asked myself if these specific traits might be part of our collective unconscious. I'm inclined to believe that such is the case, and that is my greatest motivation for writing this book. I want to share these findings so that others will have the opportunity to corroborate or rule out this possibility.

This long process of analyzing has led me to the idea that what has surfaced during the clinical situation could be part of the psychological work in store for all Puerto Ricans. At the beginning of the book, I will describe these traits, recreating the beginning of my process of analysis. Finally, I will present the concept that these traits are poles in psychological conflict, present in the collective unconscious of the Puerto Rican people. I will propose that each trait introduces an opposite behavior that constitutes a mental conflict and describes part of the mental task we face at different times in our lives.

As such, the description of my findings could further facilitate the specific study of the Puerto Rican personality. This may result in a guide that stimulates dialogue directed at a better understanding of ourselves as humans, as products of a unique historical reality in today's world.

CHAPTER ONE

Colonized Personality Disorder

Definition

L ET'S START WITH a definition of what colonized personality disorder is. The following characteristics are the bases for defining what a colonized personality is. The colonized personality and/or colonized personality traits can be found in individuals without it, constituting a psychiatric disorder in and of itself. It is constituted as a psychiatric disorder when, in addition to the presence of the colonized personality, it causes severe social and/or occupational dysfunction.

In order to diagnose colonized personality disorder, each and every one of the following criteria must be present in the individual and be associated with severe occupational as well as social dysfunction. Although a colonized personality does not constitute a disorder in and of itself, in my opinion, it is a unique finding that characterizes the development of the Puerto Rican on account of the five hundred and thirteen years of colonization that Puerto Rico has been subjected to. This experience, unique in the world, has left obvious imprints on our personalities.

The personality of each individual is the result of the interaction between biology (our genetic code, DNA) and the environment (our experiences). Experiences occur at different levels: the individual level, the social level, and the cultural-historical transformation. It is through these interactions over time that personality traits are formed. A personality trait is a predisposition to act in a predetermined, specific way.

Our experiences as Puerto Ricans have been unique and different from those of American society. Growing up in a metropolis is not the same as growing up in the oldest colony in the world.

Experiences in the colony are very different from those in the states. Knowledge of these differences is indispensable in order to understand personality traits that might develop. We will define this type of personality, taking into account these historical differences. The adversities and setbacks that we Puerto Ricans have experienced during our historical transformation are very different from those experienced by Americans.

Certainly, multiple similarities exist, but at the same time there are also differences that result from the unique situation that exists in Puerto Rico. These existential differences substantiate the presence of the personality type that we are describing here. We shall continue to describe the diagnostic criteria of, as I understand it to be, the colonized personality. These also include descriptions of the tendencies that are manifested by these individuals.

1. As a fundamental personality trait, the colonized individual presents a denial of the colonial reality of Puerto Rican territory.
2. The individual is afraid of his or her personal and economic independence.
3. This individual is dependent on and submits to the will of the United States for Puerto Rico.
4. The individual is characterized by an absence of the ideal of economic self-sufficiency.
5. This individual does not recognize his or her own work as the source of capital and pride on which he or she bases his or her self-esteem.
6. The individual presents a scornful attitude and lacks a strategic plan to achieve economic self-sufficiency.
7. This individual presents an envious and jealous attitude toward other individuals' accumulation of wealth.
8. The individual confronts problem-solving situations by creating conflict rather than promoting strategic consensus.
9. The individual's vision of the world is provincial, lacking a global and international perspective.
10. It is not an integral part of his or her education to learn and use the English language.

The description of this disorder is the starting point and the fundamental topic of discussion in this book. It refers to the description of personality traits that have evolved in Puerto Ricans during their historical reality as the oldest colony in the world. This personality has evolved over the years and has been encouraged by the historical transformation of our colonial reality. I started with the description of the pathological reality, but I will quickly move on to a description of my general findings. Based on these findings, I intend to make inferences about our personality as Puerto Ricans. Although not all these traits are present in every one of us, I think that to a greater or lesser degree, some of these traits are present in all our personalities.

I still think, however, that for us Puerto Ricans there is a struggle or conflict against these traits at some stage of our lives. In my opinion, as a human behavior professional, these personality traits are the result of our history. Moreover, they are currently the cause of the permanent nature of the colonial and territorial situation of Puerto Ricans in relation to the United States. The origin of these personality traits began with our colonial reality with Spain and has continued to develop as a response to our territorial reality in relation to the United States.

We are, and have been, for many years a colony because we contribute even with our personalities to the perpetuation of our political reality. We have direct responsibility in this colonial relationship. Personality is defined as a tendency to act in a predetermined manner. The continuation of our political reality is ingrained in our person. We are, and have been a colony, first of Spain and later on of the United States, because we act like colonized individuals.

Our colonized personality tends to make us act out and perpetuate the colonial *status*. As Puerto Ricans, we hinder and impede the process of political self-determination with our attitudes and behaviors. We behave like colonized individuals. The responsibility for our situation mainly falls on us. We must analyze ourselves and discuss how it is that each one of us, through our behaviors, contributes to the perpetuation of this situation.

The presence of colonized personality traits in each one of us is a reality that we should understand and accept. It is necessary to recognize an individual perspective of personality in order to achieve change as a nation. Historical transformation creates consequences for the individual, and these often operate on an unconscious level. This process is described as the collective unconscious.

My observations on the behavior of Puerto Ricans make me conclude that the presence of the colonized personality does not distinguish party lines. Neither does it distinguish the lines between generations, social classes, or different education levels. The absence of a marked awareness in us makes us silently complicit in relation to our colonial permanence as a nation.

In the literature on Puerto Rico that I have studied, the absence of an analysis of the psychological consequences that this long historical process of colonization has had on us stands out. This history has left profound imprints on our psychological makeup. This individual analysis is indispensable and necessary in order to be able to achieve a certain amount of social change. The understanding of the colonial personality will facilitate change, first of our personality, and eventually will bring a change in our country's *status*. The definition of our personalities will enable us to move toward collective change.

Contrary to the majority of other countries in the world, as Puerto Ricans, we describe ourselves not on the basis of personality but on the basis of our political affiliation.

The preference for the desired political *status* seems to be the popular psychology. It seems enough to know whether one is either for the Popular, Estadista, or the Independentista political parties in order to make judgments about a person. Such a definition of a person reminds me of a branding iron.

The political parties have also substituted the definition of an individual person according to their attitude toward the colonizing country at that particular time. Currently, as Puerto Ricans, we are either for the Popular, Estadista, or Independentista political parties. We are not personalities with specific psychological attributes that are a product of historical development, but rather, we are individuals characterized on the basis of the attitude that we take toward the invading city or country at a particular time.

We are either for or against Spaniards or Americans. The reality is that we are colonized personalities in conflict with our desire for individual and independent development. Our personalities are characterized by our submission to authority and our powerlessness. These traits are a product of our historical reality and our attitude when confronted by an outside power.

Our personalities have evolved during the course of the past five hundred years, during which time we have always experienced the presence of an outside power – first Spain and later on the United States of America. The control over our social behavior and our future has never resided in us, but rather perpetually in the external power of the metropolis. Our personalities manifest a dependency on external control.

In psychological terms, the locus of controlling our conduct is external. From the psychological point of view, we are dependent on the context. The external control of the nation has tinged and influenced the development of both our individual and collective psychology. Five hundred years of colonialism have conditioned us to expect the control of our behavior to come from outside ourselves.

We don't consider ourselves, or feel we are the authors of our future; rather, we see ourselves functioning in relation to the external power – that is, the metropolis. We consider ourselves good, depending on whether or not the Spanish royal court accepts us. We are good or bad depending on the attitude we assume toward the United States. The central axis from which we evaluate our behaviors is our attitude toward the invading metropolis. The definition of our self is external to us.

Our own criteria are not the locus of control. We ask ourselves how much Americans really like or hate us. We are good or bad, and the rest are good or bad based on the attitude of the invader at that particular time. This relationship with an outside power determines our behaviors. The definition of the self resides in the outside power and not in ourselves.

Our self-esteem fluctuates depending on the attitude that the metropolis assumes toward us. This is the fundamental characteristic of the colonized personality, the lack of control over one's own future. Unconsciously, and in many instances, we consciously act in such a way as to solicit approval from the external authority. Self-pride is secondary.

This lack of control has created a significant imprint and defined our individual personality. Only by acknowledging this individual reality will we be able to regain control over ourselves and create a future based on our realities, which are far removed from the colonial mentality.

Denial of the historical colonial record is one of the most primitive defense mechanisms we have as human beings. It's like the ostrich that sticks its head in the sand so that it doesn't have to face imminent danger.

A colonial state is established when a large and powerful nation imposes its presence on a smaller and weaker nation through the use of military force. On November 19, 1493, the Spanish Empire invaded Boriken, home of the Taino Indians. In 1508, Ponce de León established the first Spanish settlement on the island of Puerto Rico. On July 25, 1898, the United States Armed Forces invaded Puerto Rico as part of the so-called Spanish-American War. Puerto Rico went on to become war bounty for the United States of America. Since then, we have legally remained the nonincorporated territory of the United States. That is our reality.

Our collective future will depend on the recognition and strategic aspirations to change our colonized personality. The engine that empowers collective change resides in each one of us. It's a fight against ourselves. It is a struggle against the tendency to act like colonized individuals, as opposed to individuals in control of their individual and collective future. The point of departure exists in our personality, not in the change of status.

The main problem we face as Puerto Ricans is not that of *status*. Our political status is only a consequence of our personality. As Puerto Ricans, our struggle is against the imprint that historical transformation has left on our personalities.

We must stop behaving like colonized individuals and must feel that we hold the future in our hands. We must develop our own clear vision of our individual and collective development. The control of our behaviors should originate within us, keeping in mind the external realities, and not the other way around.

Everything that is written here is my own opinion, an opinion based on my personal experiences, as well as on my thirty-three years of psychiatric practice. I propose here that traits with strong tendencies are prevalent at all levels of our individual and collective behaviors. These psychological characteristics are prevalent in profiles of individuals who have lived in countries that have undergone a colonial process. The mental problems we suffer are deeply rooted in the cultural history of having been born in a colony. Five hundred years of colonialism cannot be suddenly erased.

When the political power of a nation resides in another nation, it creates psychological difficulties for individuals, difficulties in understanding and accepting their behavior and their reality as their own responsibility. The colonial reality persists because, consciously or unconsciously, we obediently behave like colonized individuals. Trying to maintain the appearance that we are not a colony of the United States is a denial of our political reality.

Fear of the invading power, in this case the United States of America, is what primarily awakens the defense mechanism. That is the primary gain. The dependency on the invading power is the secondary gain, psychologically speaking. As Puerto Ricans, we have to begin with understanding and recognizing that this is our reality. Our dependency on the economic power of the United States initially produces a psychological state of security. Nevertheless, the ongoing dependency eventually creates a feeling of inferiority in the human being as far as his or her capacities and abilities. Consequently, the fear of losing the benefits that one has as a colony, from what is currently considered the most powerful nation in the world, shields an individual's psychology. This fear, along with the rights to enjoy the economic benefits, encourages dependency and promotes powerlessness, thus creating a vicious circle that is difficult to break away from.

Puerto Rico, as well as many Puerto Ricans, lacks a clear vision of their own economic development, which could guide them in the direction of self-sufficiency. The dependency on political parties and federal government aid is inherent the plans of many Puerto Ricans. The development of the self, as well as human and natural resources, fails to be a priority.

The need to study, work, and grow personally through these activities becomes secondary. Like the song says, "God made work to be a punishment. I'll leave work to the ox." The worth of the individual, as well as that of a nation's, is based on the development and effective management of its resources and skills. This lack of vision for personal development characterizes the colonized personality. This also characterizes 513 years of colonization in Boriken.

CHAPTER TWO

A New Social Class Arises in Puerto Rico from the Colonized Personality: Colonized Politicians

I T IS DIFFICULT to explain, from a psychological perspective, how after 108 years of colonization by the United States, we continue to have a dismal outlook for resolving this problem in the near future. Nevertheless, we will have a better perspective if we understand the traits of the colonized mentality.

The popular notion that the colonizer is constantly exploiting the colony has not been true in the case of Puerto Rico. Although we are a nonincorporated territory in relation to the United States, we have American citizenship. As opposed to many other Caribbean nations, we can travel to and from the United States freely.

Colonial power may be observed in the political control that the United States Congress exercises in Puerto Rican affairs. All our associations and relations with the international community depend on the U.S. Congress. Commerce in and out of Puerto Rico is regulated by the United States. Tariffs are imposed on all commercial exchanges both in and out of Puerto Rico. The commercial trade is monopolistic since it is all carried out through the United States.

Unlike other American citizens, as Puerto Ricans, we have no vote in the United States Congress, and unlike other states in Congress, we have no representation in the Senate. Puerto Ricans do not have a right to vote for the president of the United States. Our national defense is controlled by the United States Armed Forces. The control of commerce and our national defense ensures the continuous control by the United

States of our territory. Puerto Rico does not have an embassy in the United States or in any other country. Neither do we have international representation, except in the International Olympics Committee and the Miss Universe beauty pageant.

As Puerto Ricans, we cannot establish diplomatic or trade alliances with other countries without the approval of the United States Congress. For the United States, trading with Puerto Rico corresponds to trading with a captive audience. As Puerto Ricans, we are barred from trying to lower costs through free trade and from competing in the global economy. Free trade competition is outside the reach of Puerto Ricans and its governors. We essentially trade only with the United States, and they are they are the ones that determine and establish prices. This monopoly is a great advantage to the United States economy.

Puerto Rico has given away this freedom to the United States. The United States economy is currently in a phase of globalization. This has resulted in greater profits for corporations that have sent their operations to countries where the costs of doing business are less. There are a great number of U.S. companies that have moved their operations to other nations, in spite of the opposition of many Americans. The proponents of these policies argue that that is the reality of the global present-day economy. They also take an antagonistic position toward economic isolationism by the United States.

Economic isolation is the best description of the Puerto Rican economy. This isolation is not out of free choice. This has been imposed by military force and the political power that Washington exercises over the Puerto Rican government. Free enterprise and competition is the cornerstone of the United States economy. Free enterprise and competition are fundamental characteristics of the capitalist system. It is precisely this capitalistic essence that is prohibited in Puerto Rico. The United States defends this capitalistic system throughout the world, except in Puerto Rico.

The Puerto Rican government cannot establish trade relations and transport its products using other competitive means, except for those approved by the U.S. Congress. The freedom to negotiate new treaties of commercial trade with the best bidder is the basis and essence of capitalism. The economy of a nation is based on these liberties – liberties that the Puerto Rican territory does not enjoy. This situation makes us the oldest colony in the modern world.

Our economy depends on the economy of the United States. The Puerto Rican economy is characterized by the dependency and subordination to U.S. control. Our economic growth and development is not based on our needs and reality rather on the needs and reality of United States society.

We can recall the transformation of our agricultural history as an example of the unilateral control exercised by the United States. For many years, the economy in Puerto Rico was based on the agricultural production of coffee, tobacco, and sugar. Then coffee and tobacco production was displaced by the export of fruits and, later on, sugar, which satisfied the compulsive needs of Americans. The production of sugar displaced other products, and sugar was primarily exported to the United States to satisfy the demand for it. Puerto Rico was not permitted to refine more than 13 percent of the sugar, whereas the rest had to be sold to Americans. U.S. companies refined this product, thereby forcing us to buy our own sugar back from them.

This monolithic agricultural production, and the incorporation of equipment and technology, displaced many farmers. Faced with this state of affairs, the United States government promoted the farmers' relocation to the United States. At that time, many Americans did not consider farmwork to be a worthy occupation. Our farmers went on to become essential laborers in the U.S. economy.

The process of Puerto Rican industrialization starts simultaneously with agricultural transformations and the massive emigration of Puerto Ricans to the United States. This process of industrialization was sponsored by the U.S. government. These policies were known as Operation Bootstrap. Our population, which until then consisted of rural inhabitants, reoriented itself to urban centers and to areas surrounding the industries that had been recently created.

Because of this, the family unit was significantly altered; it went from numerous extended families to a nuclear family prototype. This transformation from rural mentality to urban mentality changed our families and the way in which we relate to one another. In turn, these socioeconomic changes have brought about fundamental changes in our personalities.

Our best resource, the human resource, is transformed based on the designs and needs of the United States economy. The migration of Puerto Ricans to the United States began to be a daily occurrence at that time. Many of these Puerto Ricans, such as my family, also moved to New York City.

The need for human resources did not just limit itself to agricultural production; rather, it was also extended due to the needs of U.S. industries. In cities such as New York, many industries were facing serious labor-management conflicts. The unions and the power they exercised were a threat to corporate profits. Many Puerto Ricans were used as "scabs" during this process, thereby earning the hatred of many Americans. This available labor allowed for the payment of lower wages, thus it also resulted in greater corporate profits.

The need for work and the lack of it in Puerto Rico forced many to migrate to the United States to accept jobs in that economy. The Puerto Rican economy lacks self-identity; it depends on the rise and fall of the United States' economy. It oscillates, depending on changes in the U.S. economy. Puerto Rico lacks a plan for development and growth so that its human resources can be directed toward national self-sufficiency. The lack of a proper plan for Puerto Rican development and growth, and the lack of freedom to exercise free trade, are other factors that make us the oldest colony in the world.

How is this situation possible in this day and age? How can we explain how this is perpetuated? In order to answer these questions, we should take a look at the issues that have to do with the Puerto Rican personality. The personality of the individual is the product of his or her genetic code and his or her interaction with the environment.

The colonial situation in Puerto Rico has contributed to the development of the colonized personality. This personality is an adaptive response to the Puerto Rican situation and reality.

Faced with a lack of control over one's life and the impossibility for change, Puerto Ricans have developed a colonial mentality. The defense mechanisms of denial and rationalization are those that are mainly used by this type of mentality. Faced with the impossibility to change, we make use of denial to circumvent this need. Puerto Rico's colonial reality has been recognized by the United States administration of President Bill Clinton and the current president, George W. Bush. During a recent interview with a Puerto Rican politician, he was asked if he recognized that Puerto Rico was a colony. He responded that as far as he was aware, his only relation to the word colony is the cologne he uses every morning. This is a good example of the defense mechanism of denial.

Rationalization is a defense mechanism and has been commonly described with one typical example. This example relates to a fox that was seen jumping up to reach a fruit on a tree. A bear that was passing by saw the fox's frustrated attempts and asked if he needed help. The fox responded that he didn't need help because the fruit was green and that he didn't want it. Faced with the difficulties and impossibility of satisfying our needs, as individuals, we create explanations that are not based on reality. That is what the defense mechanism of rationalization means.

For Puerto Ricans, political self-determination is viewed through a colonized mentality, which frequently utilizes these two defense mechanisms. For a colonized personality, national independence is not the ideal situation. Neither is economic self-sufficiency nor the opportunity for growth and development through free trade.

The desire for economic independence and the fight against colonialism were the causes of the war between the United States and England. At that time, by defeating England's colonial power, the United States became an independent nation. For Americans, self-determination, national independence, and free trade are the characteristics on which the United States Constitution is based. For many Puerto Ricans, this similar desire for self-determination, national independence, and the opportunity to carry out free trade is the reason behind their rationalization and denial. It is paradoxical that these qualities – which are held in such high esteem by Americans, who are in conspiracy with our governors – are the same ones that deprive us of the means to achieving our goals.

The colonial state is perpetuated through the collective behavior of Puerto Ricans. In order to deal with the issue of political status and economic dependency, we have to first carry out a self-analysis of our personalities.

During this analysis, we must consider the impact that this historical reality has had on the development of our personality. The colonized personality is the product of this historical situation. Nevertheless, this should not be the ideal or the aspiration for the Puerto Rican nation. We must engender the struggle against the colonized mentality, starting with psychological defense mechanisms that are mature and efficient, in order to resolve problems. We must accept this political reality and individually contribute in order to transform it.

The economic dependency of Puerto Rico on the United States has also contributed to the emergence and development of the problem of the colonized personality. The substantial economic aid that Puerto Rico receives from the United States has promoted the colonized mentality and personality. Over the years, Puerto Rico has received billions of dollars from the United States.

This economic aid, in conjunction with a colonized personality, has made it possible for a new social class to emerge in Puerto Rico. This is a unique social class in Puerto Rico: the *colonized politicians.*

Unlike many other places, the politicians in Puerto Rico appear on the scene with the objective of essentially administrating and managing the economic aid that comes from the United States. Consistent with the colonized mentality, their objective is to use this money, thereby establishing themselves as a unique social group.

The colonized politicians have as their objective the perpetuation of their distinct social class. The colonized politician's priority is not to develop an autonomous economy that would lead to self-efficiency or to fight poverty.

Economic aid from the United States is the economic substratum that maintains them as an independent social class. They appear to be different, but all political parties share the common objective of using this aid to perpetuate dependency, and hence the colony. This social class has assumed different faces, which are used to give the appearance of a democratic process and nonauthoritarian control on behalf of the United States in Puerto Rico.

The lack of consensus among these politicians perpetuates their existence. The historical situation and reality of Puerto Ricans has been subordinate to the needs of this class of individuals. The need for national growth and development has become secondary before the needs of the groups to perpetuate themselves.

The emergence of this social class creates a semblance of democratic process in the country. Although seemingly all three political parties in existence in Puerto Rico are different, we observe traits of the colonized mentality in all of them. The political tasks in Puerto Rico extend beyond the typical task of a politician in the United States. Policy is a form of amusement here, since the power to make decisions regarding our political status does not reside with our politicians, but rather with American politicians. Our colonized politicians fight among themselves in order to be able to administer the treasury of the United States in Puerto Rico. They compete among themselves for their turn to access these funds. Access to these funds is a source of work and financial benefits for the party members. As Puerto Ricans, we describe ourselves according to the political party we belong to. In Puerto Rico, simulating alliance to a political party makes it easier to get work, and it is a group in networking that obtains financial support.

The colonized politicians serve themselves and their party and not serve the people. Politicians in the United States know how to describe themselves in terms of their attitude regarding social and institutional change. Their attitude regarding social and institutional change can be described as either conservative or liberal. The swiftness with which they favor social change is the fundamental characteristic for U.S. politicians. Conservatives are the slowest in promoting social and institutional changes; they are most resistant to change. The United States liberal politicians favor swifter changes in the social and institutional environments.

In Puerto Rico, there is no consensus among politicians as far as a change in the current colonial situation. They do not work together on the priority of change. A consensus on the need for change is indispensable in the self-determination process. It is this consensus on the need for change that will make the process of self-determination possible. Without consensus, we will not be able to change our current reality. The only outcome derived from a lack of consensus is the perpetuation and maintenance of the political social class. As far as the hypothesis that the political status quo in Puerto Rico could change, one must ask oneself, what will then happen to this new social class?

For many years, the political parties have not worked to reach a consensus that would lead us to change. Change is a threat to every political party. Faced with this threat, the parties perpetuate themselves by sustaining conflict and tribal struggle, a fight that allows the United States to maintain its colonial control over Puerto Rico.

For many years, the political parties have substituted the need to create a consensus by focusing instead on changing the administration and management of public funds. It is common knowledge that a large percentage of these funds come from the United States. It is impossible to manage our economy by exercising free trade. The lack of an autonomous strategic plan for growth and development characterizes our current reality.

The politicians in all the political parties should be making a greater effort to change this current reality. Each party should emerge with a plan for the growth and development of the Puerto Rican economy, leading to its self-sufficiency and interdependence with the international community, without excluding the United States. Opportunism and dependence on the United States have characterized the political undertaking in Puerto Rico.

It is a well-known fact that it is difficult to get work or contracts in Puerto Rico when a person doesn't belong to the political party in power at that time. Private development and small business must confront the dual negative forces of political parties and their policies. The absence of pride in self-sufficiency, love of work, and personal achievement in political parties is striking.

The lack of action, favoritism, conflict, the absence of clear definitions, the lack of an autonomous development plan, and the continual political maintenance of the status quo are what characterize the colonial parties. None of the political parties present in Puerto Rico have a clear definition of what it is that they are pursuing or how they are going to achieve it. Few are aware of the consequences that statehood, independence, or an improved, free-associated state would particularly mean for the individual.

An ardent and fervent tribal struggle is being fought without knowing exactly what is at stake. No political party has accepted this responsibility, since the colonized mentality, one might say, makes them slaves to what the United States Congress chooses. No one seems to be fighting for what is truly our biggest problem. Our main problem is how a territory of about thirty-five hundred square miles and a population of nearly four million people, with limited natural resources, will grow and develop through work as a self-sufficient nation that is interdependent on the international community, including the United States of America.

The situation in the island of Vieques is the best example of how a people and their political parties can, in joint consensus, accomplish their goal. This was the consensus that successfully changed the United States' policy and stopped its consequences, which were in the form of environmental destruction, bombardment, and new diseases that Puerto Ricans living on Nena Island had suffered.

The colonized personality exists in all political parties. This perpetuates the colonial state and dependency. Due to the generous economic aid provided by the United States, we depend primarily on them. This dependency extends itself to the dependence on political parties. In other words, the colonized personality transfers its dependency to the party. The colonized individual does not force his or her own criteria. He or she toes the party line and lives vicariously through party leaders.

The colonized personality does not exercise critical thought or question his or her leaders. No one demands a clear definition of the terms or a clear strategic plan for national self-sufficiency. The colonized personality does not demand from his or her leaders what is lacking in his or her personality, their own criteria. We stop being Puerto Ricans and instead become *estadolibristas, estadistas,* or *independentistas,* and only God knows what that means. Ignorance about the implications of each political posture is the all-encompassing reality for individuals. We focus our attention on the leader's personalities, not on the content of their messages, in part, because of a lack of analysis, motivated by the need for psychological dependence.

American society is characterized by its love for independence and its economic self-sufficiency. For them, emotional, economic, and political independence is primarily ideal. Free competition, efficiency, and international economic competition are the primary goals for Americans. Colonized individuals do not emulate these American characteristics, the lack of which does not endear us to the American government. The lack of self-sufficiency and autonomous economic development constitutes one of the greatest impediments to establish Puerto Rican statehood. The fact remains that whatever form of political status we choose for Puerto Rico, there is an urgent need for us to engage in our own economic development

In the United States, the Republican Party is the main proponent of an individual's responsibility in his or her economic self-sufficiency. In the United States, the promotion of individual wealth and the insistence that the state develop self-sufficient economies characterizes the Republican Party. Republicans have promoted, in their domestic policies, minimum federal interference in the economic development of each state. This political party emphasized independence and responsibility for economic development.

In order to grant statehood, the United States will probably demand greater economic development in Puerto Rico, as well as demand greater efficiency in the use of our resources.

Puerto Rico consists mainly of poor and indigent people, where less than half of its population is capable of work. For a great number of individuals, economic dependency and greediness have substituted industry and work.

The largest sources of work in Puerto Rico are the government and the political parties. The private sector is unfortunately the minority in Puerto Rico. In turn, the governments have made little effort to develop and strengthen private business in Puerto Rico. Colonized politicians have shown favor to big American corporations instead of small local merchant. Be it as it may, whether through statehood, independence, or a nonterritorial, free-associated state, the development of greater and more prosperous small businesses is a fundamental part of our development, particularly as we move forward toward decolonization

For the colonized mentality, what is "ours" is inferior and often associated with inefficiency and incompetence. On the other hand, what comes from the metropolis is always better. The colonized mentality holds the idea that if we do something, we will not do it well. For this mentality, free enterprise, competition, and success based on merit are absent. The order of the day for all political parties is cronyism and favoritism.

Populism is the favorite doctrine of colonized politicians. The labor laws developed by this mentality have favored the creation of permanent jobs that are not based on principles of merit. Under current government labor laws, it is almost impossible to eradicate the lack of action and mediocrity that are prevalent in the services provided by the government. A job guarantee is protected by these laws; hence, efficiency is unnecessary.

In the populist philosophy, poverty is divided among many, thus, guaranteeing votes for the party. Economic activity distributed in this manner is not conducive to the promotion of wealth. Government expansion has been characterized by the parties' need to create a base of party members. The private sector has not been systematically promoted because economic investment has been directed to promote the electoral success of the government in office at that time.

The concealed objective of populist doctrine is winning electoral votes. The increase in the number of government employees is a fact for all administrative branches. The economic promotion of the private sector is not indispensable to obtain electoral permanency.

The administration of federal funds has often been described as corrupt. The misappropriation of public funds does not distinguish party lines. Public funds and monies are often funneled for the personal gain of government administrators. This collusive attitude with corruption is prevalent in government administration.

The opportunity to take over power, and thus administer public finances, comes by fostering differences and instigating conflicts between political parties. In these last years, governmental corruption is scandalously frequent. These acts of corruption have mainly come to light through audits performed by the federal agencies. Federal audits tend to be more consistent and exhaustive. Corruption intolerance tends to be more severe in the federal government rather than in the state governments. It is easier to misappropriate local funds and not be caught.

For the colonized mentality, the motivation to participate in political activity is to guarantee the opportunity to administer public funds to one's own advantage. Political activity seems to be more about destroying the adversary than building a self-sufficient and proud society, one that gains economic benefits based on individual effort and one's own work.

Puerto Rican politics is taking a new direction. A large number of Puerto Ricans consider themselves politically savvy. The passions that political discussions instigate are closely tied with personal worth and self-esteem. It is necessary to study the colonized personality, particularly this social class of politicians, in order to make government a more effective administrator of the resources we have.

We should evaluate our politicians' behaviors and compare them to the colonized mentality that we are describing in this book.

The idea behind the title of this book, *The Governor's Suits*, refers to the recently published news stories in the Puerto Rican press alluding to the fact that the Partido Popular Democratico (PPD) recently bought the current governor, the Honorable Aníbal Acevedo Vilá, forty thousand dollars' worth of suits for his own personal use. Given the fact that this is a significant amount of money and that it doesn't come directly from the governor's work, it seemed to me to be a good example of how the colonized personality manifests itself in these politicians. This economic dependency by a party leader becomes an example of how economic dependency is for many, the ideal to pursue.

Economic dependency masks another type of dependency. I am referring here to emotional dependency. During the individual's personality development, a shift occurs from physical and emotional dependency on the mother, until later stages of maturity occur. For the colonial mentality, emotional and economic independence is not considered the ideal. The colonial mentality favors the perpetuation of dependency – a dependency on the political party and an economic dependency on the United States.

Dependence upon the party presents the incentive of greater economic rewards, which rise in accordance with hierarchical rank; for example, the governor's $40,000 suit-and-wardrobe allowance. Also quite important is the opportunity to manage the state funds and generous federal aid programs.

In order to perpetuate their control, the political parties employ strategies – including populism, an aversion to private development and a lack of interest in economic development – that would foster financial self-sufficiency. This dual dependency upon the party and the U.S. economy makes this group reluctant to change.

The political class in Puerto Rico has transformed itself into the fundamental economic unit in the Puerto Rican economy, contrary to that in the United States where small business constitutes the backbone of the country's economic development. Individualism and economic independence are promoted in the United States through a variety of incentives, including low-interest loans and tax benefits for small business.

Freedom from both government and political parties are also promoted within this system.

In Puerto Rico, this new social class has historically served to perpetuate economic dependence on the United States as a result perpetuating Puerto Rico's status as a colony.

The apparent differences in Puerto Rico's political parties seem to disappear when we examine their platforms. None possess a clear operational definition of what constitutes their preference for status. It is difficult, if not nearly impossible, to define Puerto Rican statehood, Puerto Rican independence, or the nonterritorial, free-associated state.

None of the platforms demonstrates a clear and effective plan for the creation and development of a more self-sufficient economy that would be less dependent upon the United States. Nonetheless, conflict between the parties is the order of the day in Puerto Rican politics; arriving at a consensus is the exception. In turn, this routine conflict contributes to the perpetuation of the colonial situation.

Other empires have employed the same strategy since ancient times: *divide et impera*. We continue playing the metropolis's game when we refuse to come to agreement on how to change Puerto Rico's status as territory of the United States. Absent from the political party platforms are the goals of economic self-sufficiency and the possibility to compete in the global economy.

Rather, the parties are characterized by their economic support of large U.S. corporations. They don't foster the development of small business in Puerto Rico. In Puerto Rico, the political parties are those that conduct themselves similarly to large American corporations. The government exercises too much control over the Puerto Rican economy; this control is a psychological compensation for the absence of real control.

Politicians who are dependent upon their parties are similar to those people Americans term company men. It is clear that interparty conflict stems from underlying economic problems rather than from any substantive or process-oriented issues.

Competition is focused on the opportunity to manage the financial resources of the nation of Puerto Rico – an opportunity that is made more lucrative when we include the federal funds coming from the United States. In this type of competition, party loyalty takes the place of efficacy.

It is difficult to evaluate the merits of each party's platform due to the high level of ambiguity. The political parties wait on the United States and don't define their particular prescriptions for Puerto Rico's *status*. They are all waiting for the U.S. Congress to define every proposal, and they all fear that their wishes won't be granted. The *populares* fear that the nonterritorial state won't maintain a permanent union with the United States. The *estadistas* fear that statehood won't be "Puerto Rican" enough and that such a plan would demand higher self-development and governmental efficacy. The *independentistas* fear the responsibility of creating a self-sufficient national economy.

CHAPTER THREE

The Colonized Personality

THE PERSONALITY IS the combination of physical and psychological traits that, together, make up the individual. It is the whole of behavior patterns observed throughout the life of the individual – the tendency to act in a given manner in a given situation. As individuals, we exhibit behaviors that are unique to each of us. The possible combination of personality traits is infinite. The particular combination of these traits is what makes each of us individuals.

The scientific study of personality requires the human event to be studied, specified, and defined. Personality traits are the potential individual reactions to any given situation – for example, the different reactions that an individual might express before an authority figure.

These possible reactions can be outlined through the use of a continuum of behaviors that may fluctuate anywhere between either extremes. These extremes represent opposite behaviors. For example, a positive response will be a yes; its opposite will be a negative response of no. The opposite of acceptance is rejection, the opposite of coming closer is moving away, and so on. Between these two extremes, there is an infinite number of possibilities that are distinct combinations of those extremes. This strategy facilitates the scientific study of personality quite well. This methodology has been described extensively in psychological literature. It is this methodology that I will use to describe the colonized personality. This manner of study has allowed for the design of different models. Additionally, I believe that in describing the colonized personality I have also outlined a model to facilitate its observation.

In the example mentioned above, the possible behaviors of an individual before an authority figure fluctuate between behaviors of submission and rebellion. Submission and rebellion are the two poles at either end. Between those two poles, the possible combinations are infinite. The combinations and variations of those poles make it possible to describe the reaction of any individual in the face of authority.

The more frequent a trait appears in an individual over time, the better we are able to describe it as a personality trait. Personality traits are tendencies exhibited by the individual to behave in a certain manner. Once a behavioral trait is established as part of the individual's personality, this helps us anticipate future behaviors. Through this method, we propose to describe the characteristics of the colonized personality.

Observation is the fundamental instrument of the scientific method. The scientific method is hypothetically deductive in nature. In other words, by observing events, we can formulate hypotheses and calculate the probability of an event repeating itself in the future. According to this method, repeated observation allows for the formulation of conclusions regarding an event in question. Possible conclusions can be posited for the future. In this model, a hypothesis must be grounded in observations that corroborate such an event. The hypothesis describes specific events and the relation observed between them. This model serves as a foundation and guide to the medical model. In both methodologies, systematically directed observation is the fundamental instrument.

If an observation is both repeated and corroborated by other independent observers in a consecutive fashion over time, it is promoted to the level of scientific data. The repeated and corroborated nature of these observations over the course of time is what distinguishes hypotheses, premises, and scientific laws from popular knowledge.

This type of information forms the foundation of the medical model. My psychiatric practice has been characterized by the specific, systematically directed observation of behaviors. I use specific manuals, such as the *DSM-IV*, and I routinely employ scales of behavioral observation during my professional work.

Systematic observation is necessary in order to truly understand complex concepts such as personality. Whether in studies of mental illness or in the study of personality, directed systematic observation forms the cornerstone of this process.

In terms of the concept of personality, it is the observed presence of determined behaviors in an individual on a consistent basis throughout his or her life that leads us to formulate a hypothesis – this particular manner of responding is a personality trait. The grouping of traits that manifest on a consistent basis within each individual constitutes his or her personality.

Although this may appear paradoxical, the study of pathological mental states has helped us better understand concepts, including what would be considered the normal personality of an individual. Personality traits found in many psychiatric patients are only exaggerated expressions of traits exhibited by the normal population. Not only are such traits exaggerated in the mentally ill, but these patients exhibit inflexibility and an inability to adjust to the changing nature of external reality. Pathologic traits can thus help us identify normal and abnormal patterns of behavior.

To facilitate the process of observing the colonized personality, I have designed a scale that provides a system of direct observation to identify both the presence of personality traits and their level of intensity. This will also help us direct self-analysis to detect the presence of these traits within ourselves. The use of this scale with others and ourselves will help us verify whether this personality exists and whether it is prevalent among Puerto Ricans.

The personality traits outlined in this scale were first observed in mental patients and were commonly associated with severe occupational and/or interpersonal dysfunction. In this way, they have helped define the pathological poles of behavior. Personality is defined in the context of a unique mix of opposite poles of behavior. In the present scale, I have conceived the opposite traits as the functional states of each trait.

The farther we go away from the initial pole, the easier the road is to functionality. This scale describes contrary behaviors in conflict. Conflict is at the heart of the intellectual task, as it is at the heart of personality. As individuals, we engage in an internal conflict with ourselves in determining our behavior. A behavior or a personality trait is the result of these conflicts. For each individual, it is a unique mix.

To ensure that these behaviors are expressions of the Puerto Rican personality, we can observe them in ourselves as situations of personal conflict and decision making at a given point in our lives. It will also be necessary to observe them in a consistent manner in others, in order to test the concept of this specific personality. Through this collective process, we can test whether or not the colonized personality is present in Puerto Ricans.

The scale specifies the particular behavior to be observed. It describes ten variables, each of which spans a continuum of five points. These variables describe conflicting behaviors. To my understanding, the use of such a continuum of conflicting behaviors is the best manner by which to conceptualize and describe personality traits. The testing of these concepts refers to the process of determining whether such behaviors truly exist in the Puerto Rican personality. The design and development of this scale was directed to facilitate the testing process. If valid, these traits could be replicated, and you yourself can observe them. This confirmation process refers to the reliability of the concept. As such, the colonized personality in Puerto Ricans is subject to testing and reproduction by other observers of human behavior.

These observations are the result of my personal and professional experience over the last years. It is possible that they are unique to the populations I was observing and thus don't serve to make generalizations about others. The present paper addresses this possibility. My intention in describing and making them public here is to solicit the general public in the testing process. It is a challenge for the public to either corroborate or reject them as common, valid observations of Puerto Rican behavior.

The scale is easy to use. Each trait is presented in terms of five spaces, divided equally, that describe grades of intensity of each behavior. It begins at a specific trait and extends all the way to the opposing behavior. Number 1 represents the obvious presence of the trait, and number 5 represents the obvious presence of the opposite. The different intensities and combinations of opposite behaviors lie between 1 and 5. The midpoint, 3, is the most difficult to define, since it is no longer clearly the trait or its opposite. Number 2 is the closest to the trait without approaching its opposite. Number 4 is closest to the opposite without its presence being obvious.

Each dimension represents a personality trait that forms part of my impression of the colonized personality. It is my belief that this personality is the product of 513 years of Puerto Rican colonialism stored in our collective unconscious.

The opposing traits describe the opposite of the colonized personality and represent the spectrum of conflict. In my opinion, the grouping of ten traits is what characterizes the colonized personality. Each of the ten dimensions refers to universal, human personality traits. The dimensions are described in a manner relevant to the Puerto Rican personality.

Each trait speaks of basic personality functions. The behaviors described here describe these general functions and how I have observed them to manifest in Puerto Rican subjects. Using the scale, if the total sum of the ten dimensions is closer to ten, this indicates the presence of the colonized personality in the individual. The closer the total sum of values obtained is to fifty, the farther the person is from the colonized personality.

Fifty is the highest possible score, representing a clear absence of the colonized personality. I would suggest that the use of the scale begins with self-analysis – that we first test ourselves for the colonized personality. Our capacity for self-evaluation contributes to the testing of the instrument. We should self-test in each of the ten dimensions and provide a number from 1 to 5 in each. Subsequently, add the total values obtained.

The closer the total is to ten, the more likely you are to exhibit the colonized personality, while the closer the total is to fifty, the less likely you are to exhibit this personality. Once you have done this, ask several people whom know you well to evaluate you using the scale. Calculate all the values obtained, including your own, and divide by the number of evaluators, including yourself.

Once again, the closer the result of this division to ten, the more likely you are to exhibit the colonized personality, while the closer it is to fifty, the less likely you are to exhibit the personality. Very often, we perceive ourselves quite differently from the way others perceive us.

Now we will describe each one of the scale's ten dimensions.

The **first dimension** of this scale refers to how the individual perceives Puerto Rico's legal relation to the United States. In number 1, we see a negation of Puerto Rico's reality as territory of the United States. A 2 would be an individual who denies more than accepts Puerto Rico's territorial status. Individuals scoring a 3 on this scale are ambivalent, in between acceptance and denial of Puerto Rico's territorial status. In number 4, we find persons who accept more than reject the notion of Puerto Rico as a territory of the United States. Finally, persons scoring a 5 accept Puerto Rico's territorial status.

This dimension represents the individual's skills at perceiving our historical and socioeconomic reality. Our ability to perceive both external and internal reality is referred to as principal brain function. Being in touch with reality helps us formulate a functional behavioral response to any given situation.

On the other hand, the less contact we have with reality, the greater the possibility of mental illness. The act of evaluating different realities is the most difficult and complex function for any human being. This process of analysis, evaluation, and response is unique to human beings and extremely difficult to reproduce in a computer.

Psychologists have described this basic human brain function in several ways. Russian psychologists, like Ivan Pavlov, referred to this response as cerebral orientation. They note that the primordial function of the human brain is to orient the self in time and space. With this information, the brain then executes behavioral responses.

In Puerto Ricans, this basic human function has been modified, as we have been conditioned for more than five hundred years to see reality through the metropolis in session at that particular time. At this point in time, our sense of reality is conditioned by American perceptions of Puerto Rico and what we constitute for the United States.

If we look at U.S. congressional writings on Puerto Rico or in any encyclopedia, Puerto Rico is referred to as a "nonincorporated territory of the United States."

The orienting function of the human brain has also been described as the general adaptive function of the human body. The brain tells the body to emit responses of fight or flight in new or dangerous situations. Our brains coordinate all the body functions pertaining to this act of evaluating reality. In denying reality, we postpone the decision-making process; at the same time, we postpone the best adaptive and functional responses.

Our ability to perceive our history and socioeconomic reality is a function and trait of our personality. Our consciousness of our own history and socioeconomic situation demonstrates that we are unique human beings. Just like the perception of weight, height, skin color, or eye color, perception of history is an individual personality trait.

The judicial stance of the U.S. Congress, like that under former President Bill Clinton and the current group under President George W. Bush, is that Puerto Rico is a nonincorporated territory of the United States.

The colonized personality has problems evaluating reality because reality is conditioned by the colonial situation. The colonial situation is much like that suffered by American blacks under slavery. For some blacks, slavery was not experienced as an individual reality. Many denied their reality much like those who possess the colonized personality. Freedom makes the individual responsible for his or her own future. For many people, this is a heavy burden.

As evidenced in the initial test of the first dimension, this personality function exists in all of us. Particular to Puerto Ricans, however, are the unique historic and socioeconomic circumstances. Other people have suffered and continue to suffer similar emotional situations. Americans suffered during their own colonized past, and, following that period, American blacks suffered under slavery.

This first dimension refers to the understanding an individual has of his or her own history. Denial is a defense mechanism utilized by the individual, while acceptance of Puerto Rico's colonial status, located at the opposite pole, reflects the prevailing intersubjective consensus.

Denial is one of the fundamental defense mechanisms utilized by the individual. It is also considered the most primitive and dysfunctional of all defense mechanisms. Denial does not orient the individual toward reality nor does it lead us to respond and adapt in a functional manner. The principal benefit of this mechanism is that we postpone facing reality and reacting. Characteristic of many Puerto Ricans, as described in the literature of Don Abelardo Díaz Alfaro, is the refrain "Leave it for tomorrow," the attitude of postponing things for another time.

An examination of Puerto Rican literature shows such postponement of decision making as very common among Puerto Ricans. Puerto Ricans commonly put off until tomorrow what could be done today. The popular saying "Ay bendito" (Oh my god) refers to this. The colonized do not accept their colonial reality. Implicit in this denial is the acceptance of the "status quo." Also implicit is subordination to the metropolis. The worldview of the colonized is grounded in a reality that doesn't exist and yet is nonetheless shared with others.

Subordination and fear of and dependence on the colonizer are immersed in this personality. A lack of self-confidence and an inability to form independent judgments also come into play. Underlying this personality are feelings of inferiority. Subordination and dependence demand less from the individual. Thus, it is easier to perpetuate the colony because this way of life is easier. It requires much more work to take control of our own destiny than it does to exist as a colony. It is more difficult to outline a plan for economic growth and financial independence. Denial of the colonial reality of Puerto Rico is clearly a pathologic behavior.

In addition to orienting and providing us with specific information about the world around us, the brain also provides information on our history. This information is stored in the area of our brain known as the collective unconscious.

This is composed of behavior patterns that are present from birth. These patterns arise from past experiences and are stored as reflex reactions for our survival. It has been scientifically shown that our brain can be altered and modified by life experiences. Not only the brain, but other parts of the body as well can be altered by our experiences. For example, medical knowledge surrounding sickle cell anemia shows that it arose as a means of protecting humans exposed to malaria.

I believe that within the collective unconscious of Puerto Rico, we have a need to accept or reject the colonial situation suffered during the past five hundred years. Some perceive the conflict at levels imperceptible to others. Nobody has to remind us if we are a colony or not, as this is a function of the collective unconscious. In the collective unconscious, we find behaviors learned throughout the years, which have functioned over time to work through different situations. The colonial situation of Puerto Rico is recorded in our brains.

While the choice either for or against colonialism is individual, for Puerto Ricans, the conflict is universal. Some of us would prefer to leave our colonial condition behind, but also the tendency in us is to continue this way. This behavior has been functional in many personalities through our historical development. I believe that these contrary tendencies form part of the psychosocial development of Puerto Ricans, that they are unique to us and are the result of our history as a nation.

The distinctive trait of the colonized personality is the desire for Puerto Rico to continue in its present state. This desire for continuation of the status quo can be conscious and at other times, unconscious.

The **second dimension** that we are going to examine in this scale deals with the attitude that an individual assumes regarding his or her emotional and financial independence. In number 1, we find the person who fears emotional and financial independence. In number 2, we find the individual beginning to make an effort to gain emotional and fiscal independence. In number 3, we have those displaying ambivalence regarding emotional and financial independence.

In number 4, we find those who appear to have achieved emotional and fiscal independence. Number 5 represents the emotionally and financially independent person.

This dimension is an attempt to represent something that many human development studies have described as normal personality development. In this model of personality development, we see the normal development of every individual extending from total dependence upon the mother toward an independent personality. Human life begins in an embryo that is connected to the mother through the placenta. Through this connection, the baby receives the nutrients it needs in order to develop. We breathe and eat through this connection to the mother. At birth, this cord breaks, and we begin the path to physical independence. It is this physical dependence that defines emotional dependence upon ancestors. In the early stages of life, the absence of the mother causes fear and terror of the loss of life.

As we grow and continue reaching higher levels of independence, we form our own identities. Normal human development is conceptualized as the individual who passes from total dependence to independence, to the point of self-sustainment. The final stage in the formation of the individual is the development of a unique personal identity. This model of personality development falls under the premise that body and mind are distinct manifestations of the same process of human life. The development of our physical capacities conditions our psychological development. Both developmental processes occur simultaneously, and each is the cause and result of the other.

According to this model, we begin from a dependent personality and move toward the development of a unique identity that differentiates us from our ancestors. In this developmental process, we see the acceptance of increasingly greater responsibilities in order to survive. Along the way, the individual develops his or her own judgments regarding personality. These judgments reflect a conflict between dependence on and independence from the familial criteria. Differing levels of each contested trait play out in every individual.

Dependent personality disorder, as described in *DSM-IV*, refers to individuals who are extremely dependent upon others and incapable of acting independently. Personality growth is conducive to the growth of independence. Within independence, however, there are degrees of dependency. The interdependent person recognizes his or her own sense of independence and, in turn, respects that of others. The mature individual takes responsibility for his or her life and welfare. This person accepts that we are who we are based on conscious choice. Life and personality are an exercise of the individual's free will. This individual does not blame others for his or her misfortunes and bad decisions. Life is a continuous process of trial and error. By making mistakes, we learn and develop as humans. This is the nature of human life.

The more dependent the individual is, the greater the tendency to blame others for his or her own mistakes. For some, we are who we are because our parents, friends, or partners made us this way. Implicit in this tendency are fear of independence and a lack of personal responsibility. The independent person doesn't exclude work from interdependence but is guided by clear guidelines of social justice. Collaborating without resorting to exploitation reflects a respect for freedom of choice. The interdependent person respects his or her own as well as others' right to their own opinions.

In turn, dependence on the metropolis has impeded the normal process of personality development among Puerto Ricans. Without the colonizer, we question our own existence and survival. The colonized personality is immature and is predisposed toward dependence. This personality holds the United States solely responsible for our present colonial situation. The higher the degree of dependency, the greater the tendency to hold others responsible. The pathology of these extreme personality tendencies is also described in the *DSM-IV* manual of mental disorders as antisocial personality disorder.

The colonized personality is characterized by a fear of independence and an inability to work with individuals who have their own, distinct criteria. The colonized personality is incomplete, insecure, and fearful of those who are independent.

In the **third dimension** of the scale, we will examine the individual's attitude regarding the United States' objectives for Puerto Rico. Number 1 represents the individuals who depend on and are subservient to the United States' wishes for Puerto Rico. At number 2, we find individuals who begin to criticize the United States' objectives for Puerto Rico. The individuals in number 3 feel ambivalence regarding the United States' objectives toward Puerto Rico. In number 4, we see the individual is critical but still dependent and subservient to the United States' position on Puerto Rico. In number 5, the individuals make their own judgments of the United States' attitude toward Puerto Rico.

Reliance on outside opinion, in this case that of the colonizer's, has been described in the psychological literature as the locus of control. Within this psychological variable, two fundamental types of persons are described. One is a person who maintains control of his or her internal behavior. For the other type, control is exercised over external behavior. An example of the latter is the person who decides what to eat after seeing an ad on television rather than when he or she is hungry. The individual whose locus of control is internal eats when he or she is hungry. This fundamental characteristic of individual personality has also been defined in psychological and operational terms as independent or dependent on external context. Psychological tests have been developed to gauge these characteristics.

In one of these tests, the individual takes a test that asks him or her to read a message out loud to determine what type of errors and how many errors are made during the reading. For example, the exam shows the word "red" written on a screen; however, the word is written in blue. The more errors the individual commits due to these interferences, the greater the extent to which their responses are affected by context. The colonized personality is dependent on context, and the control of behavior is focused externally. The colonized personality is both very impressionable and insecure.

The current colonial situation, under which judicial power resides in the U.S. Congress, has contributed to the formation and perpetuation of this personality trait in Puerto Ricans, but its origins date back to the era of Spanish colonialism. This personality trait deals with the capacity to separate the whole from its parts. It also refers to the ability to form a gestalt of a situation without being unduly distracted by the various elements that come into play.

The **fourth dimension** describes the individual's attitude toward his or her financial independence. In number 1, it is the person who is not financially self-sufficient. Number 2 describes the person who begins to accept the necessity of financial independence. The individual in number 3 is ambivalent about financial independence. In number 4, it describes those who are on the path toward financial independence. Number 5 describes the financially independent person.

This dimension examines an idealistic versus materialistic philosophy of life. According to the materialistic vision, mind and body are manifestations of the same phenomenon – that of human life. The idealist visualizes the mind and the body as two distinct entities that manifest in parallel but independent forms. According to the materialistic vision, society and economy are manifestations of a singular phenomenon, human social life. The idealist sees the development of society as independent from economy. For the materialist, individual development, much like nation-state development, is intimately linked to economic development.

Individual economic development is fundamental to the personality in a materialistic philosophy. Development of a proper economy is fundamental to a culture's sociological development. Culture and economy are intimately linked in the materialist perspective. For the idealist, culture and national economy are two very different things. The colonized personality tends toward idealism in the manner described above. For the colonized individual, the development of an economy conducive to self-sufficiency is not a priority.

The idealist separates ideas from things. The materialist sees ideas and the brain as different expressions of the same thing: life. Culture and economy are both expressions of a nation's existence. The economy provides a foundation for the individual in the same way that it provides a foundation for the nation-state. Countries thrive in proportion to their economic development. Culture is the collection of particular traits that describe the production and management of material goods necessary to survive.

The ideal of developing a self-sufficient economy reflects a materialist vision of life. To defend culture, without developing financial independence, constitutes an idealistic vision of reality. A country and its culture exist within the context of economic development. The absence of a plan to develop financial independence implies both acceptance of and a desire to perpetuate Puerto Rico's colonial status. Without developing a self-sufficient economy in Puerto Rico, we cannot talk about a nation that is mature and cognizant of its basic responsibilities. Those who don't see the need for Puerto Rico to develop financial independence are behaving like faithful colonial subjects.

The **fifth dimension** of personality that we will examine describes the relation the individual establishes between work and self-esteem. In number 1, we find the individual for whom work is not essential to self-esteem. Number 2 describes the individual that begins to establish a relationship between work and self-esteem. Individuals who are ambivalent toward the relationship between their work and self-esteem fall under number 3. Number 4 represents those individuals who are in the process of making work the essence of their self-esteem. Number 5 describes those individuals whose work is the essence of their self-esteem.

This dimension of the personality is particularly important for the colonized individual. Colonizers use the work done by the colonized for their own benefit. During the Spanish period, the work of the Taino Indians, and subsequently that of the black slaves, enriched the Spanish empire, but it did not enrich the workers. This economic control over production by the metropolis tends to create a mental attitude where work is not seen as a personal benefit. As a consequence, an attitude of resentment is inappropriately transferred toward work. The problem then stops from being the colonizers and becomes work itself.

This shift of resentment toward work is prevalent in Puerto Rico, both within government agencies and in private companies. A customer service attitude is not the ideal to follow for many Puerto Ricans; they see "attitude" as that which is symbolized by the popular saying "The brute lives off of his work, while the wise man lives off of the brute's work".

Negative attitude toward work is not unique to the colony, but rather is universal in all social groups. In order to scientifically understand human behavior, we need to look at the physical laws of matter. After all, human beings are made up of matter. Work, within the framework of the laws of physics, can be described as "the energy used in carrying matter through certain distances at specific velocities." In simple words, work implies the use of energy to advance the movement of matter, our bodies, and other objects. One of the laws of physics states that bodies at rest tend to remain at rest. Working takes work and an implicit desire to struggle against the laws of gravity. Working isn't easy from the perspective of physics.

This dimension tries to explore the consequences that the colonial situation has had on the physical processes of those individuals, especially their attitude toward work. Within the structural sociological perspective, individual work is the source of all capital. The awareness each individual has of this reality is variable. Many people refer to this dimension when they talk of work ethics. From the physical and medical perspective, the most important quality our bodies have is the ability to produce work. It is this ability that distinguishes life from death.

The contempt toward one's own work is an alienation and denial of existence. Our bodies and their ability to generate physical and intellectual work are the most precious possessions we have. We need to learn to differentiate between an attitude toward work brought about by five hundred years of colonization and one that is healthy, which we must adopt.

The labor force in Puerto Rico works halfheartedly. This is mainly due to the multiple aid programs offered to Puerto Ricans, which require that individuals be unemployed in order to be eligible. If this aid were to include a "work" component, then our national production would become much larger. Assuming that half of the labor force is currently working and not receiving SSI (Supplementary Disability Insurance) in federal aid, I would speculate that if they were to receive this aid, only a third of them would keep on working. Here in Puerto Rico, disability aid is given through social security, and it is paid for by the recipients during the years they have worked. In the United States, however, one need not have paid social security or the necessary trimesters in order to qualify for disability benefits. These benefits are used to fight poverty.

In a country like ours, where poverty levels are high, the outlays that the United States has to send Puerto Rico, in terms of the SSI, are astronomical. This has thereby created a nation of individuals disabled from work. Financial aid to fight poverty should be handled through work incentives so that work can be seen as the main source of capital and pride for an individual's self-esteem.

The **sixth dimension** of the colonized personality evaluation scale refers to one's attitude toward economic planning. In number 1, we have those individuals for whom there is no economic growth plan. In number 2, the individual starts to acknowledge the need for economic planning. In number 3, the individual is ambivalent toward economic planning. In number 4, the individual starts to become active in an economic development plan. In number 5, the individual is active in an economic development plan. This dimension explores the importance and priority the individual gives to the process of planning.

This dimension refers to the personal ability of looking ahead and being able to plan based on one's expectations and resources. Living in the present is a common characteristic of primitive societies; as societies develop, however, they begin to acquire both a dimension of time and the ability to think ahead. Lack of planning, on the other hand, is more common in developing countries and in groups pertaining to low socioeconomic levels. The colonized personality looks down on planning; thus, he or she lacks a strategic plan for achieving economic self-sufficiency.

In the **seventh dimension,** the scale attempts to measure an individuals' attitude toward private property and other people's accumulation of capital. In number 1, the individual is envious and jealous of other people's private property and accumulation of capital. In number 2, the individual begins to acknowledge the need for respecting private property and being competitive with other people in the accumulation of capital. In number 3, the individual is ambivalent between an attitude of jealousy and envy and the need to respect and to be competitive in regards to private property and the accumulation of capital. In number 4, the individual starts to be competitive regarding the accumulation of capital. Number 5 describes the individual who respects private property and is competitive regarding the accumulation of capital.

In early human civilizations, common property was the law, and available resources belonged to the strongest and to the most aggressive individuals. Private property was inexistent. Hence, war and fighting were daily events needed for satisfying life's needs. The strongest male kept all the females, and the more hostile tribes destroyed the weaker ones, taking over their resources and women. This type of behavior characterizes the animal kingdom, but as societies evolve, the laws of nature start to become modified by those of social justice. As this type of social order becomes more defined, private property finally emerges.

The attitude individuals have regarding private property also forms part of the collective unconscious. The most primitive reaction toward this situation is that of jealousy and envy toward other people's acquisition of capital. The mature attitude toward this situation is accepting private property and becoming competitive in relation to the acquisition of capital. The person with colonized attitude is dependent upon what others have rather than work on acquiring capital for oneself.

This primitivism also characterizes the behavior of gangs and mobs in Puerto Rico. Crime in Puerto Rico, which is very high and rampant, is originated by this colonized personality. I remember reading the Interpol statistics of crime in Puerto Rico, according to which, we have the fifth highest rate of assassination per capita in the world.

It is not hard to see that this problem is multifaceted, but we should not discard, however, the possible relation it has to the colonized personality. I was happy to read in the newspaper that the Puerto Rican chief of police understood that there was a problem of mental health behind these figures. I agree very much with the chief's analysis. The damages made by this historic colonization process in Puerto Rico are incalculable. As long as the colonized personality predominates in our society, these damages will be hard to amend. The reason the colonized personality lacks a sense of property is that, in reality, we have never owned Puerto Rico.

The **eighth dimension** describes the individual's style toward problem solving. Number 1 describes the individual who confronts problems by creating conflicts. Number 2 is the individual who begins to see the need to create strategic consensus for problem solving. In number 3, we have the individual who is ambivalent in his or her style of solving problems. Number 4 describes the individual for whom problem solving through looking for strategic consensus predominates. In number 5, we see the individual who confronts problems by looking for strategic consensus.

Problem solving requires consensus building: initially when defining the problem, and then in the methodology used for solving it. One way of denying that a problem exists is through creating conflicts that divert one's attention from the problem. The conflict is inherent to the problem, and that is why it is constituted in the problem. The purpose of consensus is to identify the nature of the conflict and subsequently solve it. Creating additional conflicts to those already inherent to the problem does not lead toward its solution, but rather to its aggravation.

The prevailing style of the colonized personality is that of creating conflicts while pretending to work. This has characterized the behavior of Puerto Rican political parties. As soon as one of them says yes, the other says no without examining the merits of the situation.

Human nature is characterized by the fact that events are perceived differently among individuals. If an old lady falls down in the marketplace, for example, and we ask ten individuals who were present, what happened? each of them will give us a different version. This is human nature; difference and variability are the laws of the human evolutionary development. Refusing to accept this reality prevents a person from confronting the solving of problems in a manner that is competent.

As the saying goes, "There are many ways to peel an onion" or, said otherwise, "Many roads lead to Rome." Accepting our differences leads us toward strategic consensus building, which, in the long run, will solve the conflictive situation. The territorial situation in Puerto Rico will not be solved unless the political parties abandon their tribal struggle. The fight against poverty requires strategic consensus among all the parties. Dependency and the poor Puerto Rican economic development also requires strategic consensus on behalf of all the parties.

The recent economic crisis, the temporary closing of government, and the need for the churches to intervene in state matters make this style of creating conflict self-evident. The only ones who benefit from this style are the political parties, who justify their existence thereby as a social class and as dependents upon the permanence of the colonial state of Puerto Rico. The survival of these colonized parties is threatened by the possibility of a unilateral solution by the United States to the colonial status of Puerto Rico. This unilateral solution cannot be discarded in the face of the loss of political credibility that the United States is suffering from as a consequence of the war in Iraq.

The **ninth dimension** relates to the type of worldview an individual has. Number 1 describes the individual who has a provincial worldview. Number 2 is the individual with an insular worldview. Number 3 describes the individual who views the world as consisting of Puerto Rico and the United States. In number 4, the individual has a worldview that goes beyond Puerto Rico and the United States. In number 5, the individual has an international and global worldview.

This dimension tries to evaluate the individual's perspective of the world around us. The mature personality sees beyond itself; the immature personality, in turn, is egocentric and narcissistic. We are all part of the human race, and as such, we are confronted with the same basic problems of life. Survival in a world where we grow as humans in disproportion to the growth of natural resources is the common denominator. Societies can be distinguished by their variations in how they manage these problems common to us all. No one is better than anyone else.

Every culture can learn about those differences with other cultures. Humanism entails looking beyond one's own social order. As we develop, global worldview becomes more necessary. In our times, the view of a global economy is necessary for individual and national survival.

The **tenth dimension** within our scale involves the importance given by the individual to the use and learning of English. Number 1 represents the individual who refuses to use and to learn English. In number 2, the individual accepts the need of using and learning English. In number 3, the individual is ambivalent toward the use and learning of English. In number 4, the individual begins to use English and to make it part of his or her learning. In number 5, the use and learning of English is an integral part of the individual's education.

Language is the means human beings have for communicating with one another. Communication has a dual function: it has an expressive dimension, and it has another dimension directed toward modifying other people's behavior. The former describes the function of letting others know about us, our ideas, our emotions, and our perceptions.

The latter refers to what has been labeled instrumental value of language – that is, its ability to modify the behavior of others. Language has a common biological substratum in all human beings. This was described by Noam Chomsky as an inherited grammar. This concept implies that every human being has the ability to learn any of the languages spoken by humans.

For Puerto Ricans, English is very important for several reasons. Primarily, it is the preferred language in the U.S. metropolis. If we wish to be able to modify the behavior of Americans toward us, we need to be able to communicate efficiently in their language. Second and most importantly, English is the language mostly used for doing business in the present global economy.

Summary

Dimension 1 concerns the level of contact individuals have with the Puerto Rican historical and socioeconomic reality.

Dimension 2 concerns the personality's level of maturity.

Dimension 3 concerns the relationship with authority figures. It describes the degree of submission and criticism in the face of the metropolis, making reference to the degree of dependency upon external criteria and an external locus of behavior control.

Dimension 4 concerns individuals' choice of life philosophy: idealistic versus materialistic.

Dimension 5 concerns individuals' attitudes toward their job.

Dimension 6 concerns the importance and priority given to the process of planning. This dimension explores one's individual orientation regarding time.

Dimension 7 concerns the level of acknowledgment and respect given to private property. It also describes individuals' attitude toward the accumulation of capital by others.

Dimension 8 concerns problem-solving style. It also describes leadership style within the social group: authoritarian versus democratic.

Dimension 9 concerns the individuals' perspective and vision of the world community.

Dimension 10 concerns the use and learning of English. It further describes the individuals' level of acceptance of the dynamic nature of the language.

The individual with a colonized personality manifests the following characteristics:

1. Wants the continuation of the *status quo* – the continuation of the U.S. territorial colony
2. Has an immature, dependent, and narcissistic personality and depends emotionally and economically on others; has a magical and primitive thinking; and possesses the agricultural mentality of the parasite
3. Is subordinate toward authority; is not critical toward the power of the U.S. metropolis; and depends on external criteria, with an external locus of behavior control
4. Has an idealistic philosophical perspective of life, in which mind and body and culture and economy are distinct and independent entities
5. Does not realize that body and work are the most precious possessions one can base one's self-esteem on
6. Lives in the present and does not plan for the future
7. Has no respect for private property and shows envy and jealousy toward other people's accumulation of capital
8. Is authoritarian and faces problems and social situation by creating conflicts; is egocentric and considers collective consensus to be irrelevant
9. Has a limited and narrow view of the world and does not see beyond his or her immediate reality
10. Has a static definition of language and does not acknowledge the need to include English learning in their education

The individual with a noncolonized personality manifests the following characteristics:

1. Wishes to change the territorial *status* of Puerto Rico with regard to the United States and acknowledges the Puerto Rican nationality
2. Has a mature personality, is emotionally and economically independent, is interdependent upon others, and has superior and functional defense mechanisms
3. Is critical toward authority figures, has his or her own criteria but is flexible and rational toward the metropolis, has an internal locus of behavior control, and is independent of his or her context
4. Has a materialistic philosophical perspective of life and visualizes mind and body and culture and economy as distinct manifestations of human life and societies respectively

5. Values body and work as the most precious possessions on which to base his or her self-esteem
6. Lives in the present, while considering the future, and plans life in accordance with his or her expectations and resources
7. Respects private property and reacts in a competitive manner toward the accumulation of capital
8. Has a democratic leadership style and promotes consensus and visualizes collective consensus as the style of national maturity
9. Has a global and international world vision
10. Uses and makes English an essential part of his or her education and considers the language as a dynamic entity

Colonized Personality Evaluation Scale

1	2	3	4	5
Denies the territorial reality of Puerto Rico	Mostly denies rather than accepts the territorial reality	Ambivalent toward the territorial reality	Largely accepts rather than denies the territorial relationship	Accepts the territorial reality

1	2	3	4	5
Emotionally and economically dependent	Starts to become emotionally and economically independent	Ambivalent toward emotional and economic independence	In the process of becoming emotionally and economically independent	Emotionally and economically independent

1	2	3	4	5
Dependent and submissive toward the United States	Starts to become critical of the United States	Ambivalent toward the United States	Starts to have own criteria regarding the United States	Has own criteria regarding the United States

1	2	3	4	5
Not economically self-sufficient	Accepts the need for economic self-sufficiency	Ambivalent toward economic self-sufficiency	In the process of becoming self-sufficient	Economically self-sufficient

1	2	3	4	5
Does not see work as the essence of one's self-esteem	Begins to form a relation between work and self-esteem	Ambivalent between work and self-esteem	In the process of making work the essence of his or her self-esteem	Considers work as the essence of one's self-esteem

1	2	3	4	5
Has no plan of economic development	Acknowledges the need for economic planning	Is ambivalent toward economic planning	Initiates economic planning	Active in economic development planning

1	2	3	4	5
Envious and jealous of the accumulation of capital by others	Acknowledges the need for competitiveness	Ambivalent toward the accumulation of capital	Starts to become competitive in relation to the accumulation of capital	Competitive in relation to the accumulation of capital

1	2	3	4	5
Creates conflict when facing problems	Sees the need for a consensus when facing problems	Ambivalent toward problem solving	Sometimes looks for consensus when solving problems	Primarily looks for consensus when solving problems

1	2	3	4	5
Has a provincial worldview	Has an insular worldview	Sees the world as consisting of Puerto Rico and United States	Sees the world as more than Puerto Rico and United States	Has an international and global worldview

1	2	3	4	5
Rejects the use and learning of English	Accepts the need of English	Ambivalent toward English	Starts to use and learn English	Makes the use and learning of English essential in his or her education

CHAPTER FOUR

Beyond the Colonized Personality

O VER THE COURSE of this work, I have tried to share with you my experiences and ideas relating to the Puerto Rican personality. My concern and inquiry on this issue began during my years as a student at the University of Puerto Rico. Although I was studying in the School of Natural Sciences, I took a large number of elective courses at the School of Social Sciences. Some of these courses, especially in the department of psychology, were taken between 1968 and 1969.

It was during a class in abnormal psychology, which I took with Dr. Carlos Albizu, that we discussed issues concerning what characterizes the normal personality. We also discussed during the course the different theories relating to the normal personality of an individual. One day, during one of his classes, Dr. Albizu asked us our opinion about the characteristics of the Puerto Rican personality. My answer at that moment was that it was an impossible task. At that time, I believed that there was no normal Puerto Rican personality; rather the contrary, there were many Puerto Ricans with different personalities.

During that course of abnormal psychology, I learned that "normal" referred to a statistical concept. "Normal" refers to the frequency in which certain events occur. Natural events are normally distributed on a curve of probabilities; the most frequent events on the curve describe what is regarded as normal.

The human personality is a natural event, and it can also be represented in this type of distribution. In this work, I have tried to define the personality traits that most frequently occur among Puerto Ricans. In my opinion, the personality traits described here describe a significant number of our population, and they are particularly unique among us.

I believe that the colonized personality is a reality among the general Puerto Rican population. The tendencies to act as colonized individuals may hinder Puerto Rico's process of decolonization. A clear awareness of how our personality contributes to the perpetuation of the colonized personality can help us understand why the colonial *status* has prevailed in Puerto Rico.

During my years of psychiatric practice, the question of the Puerto Rican personality has been persistently present in my mind. Although I still maintain that there is not one single Puerto Rican personality, but rather that there are several different personalities, I am convinced now that the colonized personality is common and normal among Puerto Ricans.

Based on my experiences, this personality is very common among the population I have examined over the past years. My reluctance to characterize what is "normal" has changed, thanks to the experiences I have had in the medical field. My medical training has taught me that there are traits and characteristics that can help a physician tell the difference between a state of health and that of an illness. My resistance to classify people, and in a sense to typecast them into specific categories, has gradually dissipated as I have realized how useful this classification system is in treating my patients. Being able to correctly classify different states has oftentimes made the difference between life and death; likewise, distinguishing between simple indigestion and appendicitis has allowed me to take my patients to an operating room rather than their grave.

Scientific classification has many limitations, but it also has quite a few advantages. While it is true that there isn't a single human being, but rather many different ones, classifying individuals allows us to distinguish them from a horse or from an ostrich.

When we classify any event, we take a step away from reality so that it can be studied and distinguished from other events. The scientific method allows us to determine the exact frequency that an event occurs compared to other events on the curve of normal distribution. For example, how many extremities does a spider have compared to a horse or a human being? The relative occurrence of each trait helps us distinguish those groups that have similar occurrences from groups that have different ones. But if the colonized personality clearly does not have a discernable existence, then why write about something that does not exist?

I believe there are many reasons to do so. To begin with, the scientific study of the human being has allowed me to help thousands of patients during my years of medical practice. These efforts have also helped me to understand myself better as a human being. It has also enabled me to understand that being a Puerto Rican is something unique. Being Puerto Rican is not the same as being Spanish or American. Although we are all part of the human race, we are all different, since our socioeconomic and historic circumstances are different. Being different does not make us better or worse; it doesn't make us inferior or superior.

Variability in nature is the essence of survival. It is important to understand what this variability in Puerto Ricans consists of, given that this knowledge will allow us to create a precise account of what we can rely on for our survival. Thus, this account of our personality can enable us to better adapt to our present circumstances.

Five hundred years of colonization is a historic reality for Puerto Rico. These experiences, accumulated over time, have conditioned our behavior. Colonization has had direct effects and consequences on our personalities. The scientific study of these effects is not only necessary, but it is the responsibility of all experts on human behavior.

Our personality as Puerto Ricans has suffered severe damages, and it is necessary to define them and correct them. Accepting the pathology that this process has created in our personalities will be the first step on the way to mental health. This path will become effortless as each person begins to define what is necessary for change and the direction of this change. That has been my intention in writing this book, as well as in creating the colonized personality evaluation scale. I have been a psychiatrist for thirty years, and I cannot stop thinking as one. It is my profession, and I feel very proud to practice it.

Puerto Rico has a serious mental health problem, and in my opinion, this problem is based on our colonized personality. The road to mental health for Puerto Ricans must guide us from the colonized personality to the noncolonized personality.

The noncolonized personality is observed in an individual who accepts the colonized and the territorial reality of Puerto Rico. Furthermore, he or she does not wish to be part of the United States territory. The noncolonized individual visualizes statehood, independence, and the nonterritorial free-associated state as noncolonial alternatives. As a U.S. state, we would have greater influence and control over our affairs in Congress and direct access to presidential elections in the United States.

In addition, each state has certain autonomy from the federal government. Consequently, the road to independence would offer us complete control and responsibility of our own future as a nation. As a nonterritorial free-associated state, we would be freed from unlimited control by the United States Congress and thus allow ourselves to establish a relationship of mutual accord with the United States.

The colonized personality shares the agricultural mentality of a sharecropper living and cultivating the land of the deputy. We have a constitution, and we have control over Puerto Rican internal affairs. However, we are a territory of the United States; hence, we do not have possession of Puerto Rico. The noncolonized personality seeks to change this sharecropper relationship into one of either partial ownership (partner) or absolute ownership. The concept of private property is very clear to the noncolonized personality. Those who do not know their history are likely to repeat their past mistakes.

The colonized personality is emotionally dependent and insecure, whereas the noncolonized is emotionally independent and self-confident. The colonized is characterized as being immature and dependent on others. Initially, he or she depends on his or her parents and family and later transfers that dependency to others. As we've discussed previously, this transference for many is directed toward their political party. In the colonial mentality, dependency is primarily transferred to the metropolis.

The dependent individual needs an external power in order for him or her to acquire some level of control over their behavior. This kind of mentality demonstrates a great need for external controls when guiding their life. The immaturity of their personality does not allow them to trust themselves or their instincts. Their dependency makes them doubt their personal value. They look for others to take on the responsibility for their own actions. Life frightens them; they are afraid of loneliness. They establish relationships with others based on their needs, not on the genuine desire to share responsibilities.

The colonial situation has had an important effect on the normal psychological development of an emotionally independent Puerto Rican. Normal personality development takes place in the context of an environment that compensates independent activity. The circumstances of a colony, however, create the contrary by reinforcing dependency.

Emotional independence is the mature state of an individual's personality. We are Puerto Ricans only for ourselves, the International Olympic Committee, and the Miss Universe pageant committee. Before the United States and the rest of the international community, we do not have an independent identity. We don't have ambassadors in the United States or in any other country in the international community. The maturity of what would constitute a Puerto Rican identity has not been reached nor accepted by other countries. The Puerto Rican identity is in its developing phase, given that the colonized personality predominates among Puerto Ricans. The dependency of the colonized personality results in our immature nationality.

The colonized personality is egocentric and narcissistic; for him or her, nationality is not a collective act. A colonized people do not recognize the collective process as the mature state of selfhood; therefore, they can only maintain unilateral relationships in which differences are not respected. A single act of personal aspiration changes the colonized individual's reality. They have thoughts which are magical, and even grand, in nature.

The process of collective convictions is substituted for personal aspirations. The colonized personality does not care for the process of collective maturity and does not seek consensus. Whereas the essence of a democracy is the thought of the majority for the colonized individual, this is substituted for a constant need for recognition. They constantly look for recognition, and they need external approval or rejection in order for them to feel that they exist.

They create conflict just to reaffirm their own existence and value. Their value does not come from within or from their own merits. They acknowledge neither consensus nor the collective process as a path that leads toward maturity. If others do not accept their position, they simply render the fact as irrelevant. Colonized individuals feel that they hold the truth. They do not accept being subjected to criticism; criticism troubles them. They consider their positions to be unquestionable. Dialogue and exchange only occur with those who accept them blindly.

The colonized personality does not see beyond his or her own self, and they do not acknowledge the need for consensus. Like babies, these individuals believe that the more they yell the more they will get what they want. They delegate their own decision making to the United States Congress. We have been a territory of the United States for 108 years, and at no time during this period have we said no, in a *national consensus*, to this condition.

The process of decolonization will only be achieved through developing mature personalities. The mature personality acknowledges the need for consensus, dialogue, and criticisms as necessary steps toward self-improvement and change. The mature personality accepts the development of consensus as something indispensable for change, and the path toward the maturity of the personality. A consensus on the need for change will propel Puerto Rican identity to other levels of maturity, thus, surpassing individual necessity with collective well-being.

A Puerto Rican identity may exist in our minds, but its existence will not be acknowledged in the United States, or in any other country, until it is expressed as a *consensus*. Submission and dependency on the United States has no place in a mature, noncolonized personality. The United States made us a territory by force. They have also unilaterally given us American citizenship, and similarly, they may unilaterally take it away. Colonized individuals do not have their own criteria, their own will, or the ability to question the metropolis.

It is true that as Puerto Ricans we are American citizens. It is also true that the U.S. Supreme Court has acknowledged Puerto Ricans' right to decide on internal legal issues. An example is the case of the *pivasos* or mixed votes – the votes that gave victory to the current governor of Puerto Rico during the last elections. These votes were questioned by different political parties, each with its own particular interpretations, and were adjudicated in local courts although the matter was eventually taken to federal court. Nevertheless, it is not true that as Puerto Ricans we are in possession of Puerto Rico.

Only the United States Congress has the jurisdiction to change this situation. Colonized individuals believe everything their leaders tell them. They do not have criteria of their own, and they are subservient. They have a subservient mentality both toward the United States and their political party. Thus, they depend on the will of others. If the party leader states that we are a free state, they believe it; if they are told that our American citizenship is permanent, they also believe it without questioning it. They do not realize that the United States has the jurisdiction to unilaterally take away our American citizenship. If a party leader tells them that statehood is just around the corner, they believe that as well. It is sufficient enough for the party leader to say that we can be independent for them to believe it without questioning it. How is that possible?

Colonized individuals lack critical thinking. For them, submission and dependency substitute analysis and critical thinking. Colonized individuals criticize merely for the sake of criticizing. For this mentality, the purpose of critical thinking is not to search for the truth or explain the facts. For the colonized mind, facts are accepted as acts of faith rather than as acts of reason. Reason and critical thinking are taken negatively, and submission substitutes reason.

The colonized personality is afraid to criticize the United States. Criticizing the United States does not occur to them simply out of fear of losing economic benefits. As long as money is flowing in, colonized individuals do not care that the United States has the ownership rights of Puerto Rico.

The sharecropper did not criticize the landowner because he lived and nourished off his farm. This personality is insecure and fights for things that it does not understand. It struggles without knowing what it is fighting for.

Given the imperialistic nature of the United States, it has held and kept the present territorial situation. The United States is a militaristic nation that exercises its power based on its military and economic power. Who would find candy bitter? Why would the United States give up their ownership of Puerto Rico? This matter is not discussed in the United States since such a discussion would eventually force negotiations. Puerto Ricans fear negotiations because they are aware of what is at stake. The noncolonized personality has to assume responsibility for criticizing the leadership.

It is the responsibility of U.S. citizens to criticize their politicians when they negotiate issues relating to Puerto Rico. Neither the United States in general nor its citizens should feel proud of possessing the oldest colony in the world. Americans should make their position known to their politicians, whether or not they want to perpetuate their possession of Puerto Rico.

As Puerto Ricans, we must be critical of our political leaders. We need to be clear on the points of negotiation. Autonomy, along with permanent union, is analogous to a pipe dream. Two nations should be free to negotiate the terms of their relationship, and as in any contract, these terms should be subject to change. Where is the freedom in this associated state, while we continue to be a nonincorporated territory of the United States, without an identity of our own on the international stage?

What would be the official language in Puerto Rican governmental transactions should statehood occur? Likewise, what would happen to our participation in the Olympic Games? More importantly, what additional federal tax contributions would Puerto Ricans have to make in statehood?

Permanent citizenship of the United States is only possible if we become a state. Only then would it be a constitutional right rather than a right by decree of the United States Congress.

What would be the basis of the Puerto Rican economy if we were to become independent? Would that independence require relinquishing all the benefits we have accumulated during the colonial period of the past 108 years?

It is necessary for the colonized personality to abandon subservience and to substitute it with critical thinking. Many heads are better than one. It is necessary for political leaders to stop behaving as if they were caudillos and accept criticism as part of the constructive process. Only in a critical and evaluative process will we be able to define our needs and our relationship with the metropolis.

The United States, the Puerto Rican government, and the political leaders have the responsibility to incorporate the Puerto Rican people into this critical dialogue, which in turn would allow us to define our realities, our needs, and the methodology necessary to improve.

As Puerto Ricans, we cannot allow history to repeat itself and continue with a relationship that is imposed on us, which does not lead toward growth and the development of our nationality and economy. We need to begin by defining what it is that we want and how we are going to achieve it. Only in this way will we be able to negotiate in an intelligent manner.

The nation of Puerto Rico cannot afford to delegate this historic responsibility to their political leaders. After all, they have not been able to do this during the last 108 years. In addition, their record on the recent economic crisis makes it obvious that they are not competent in dealing with internal affairs.

The traditional attitudes of these governmental administrations have been that of spending more than they have. The structural deficit in the government administration is a product of the colonized mentality. The colonized do not value an autonomous economy. They choose to rely on credit, thus, leaving us to drown in debt, disregarding both future generations and our survival as a nation. The government's structured deficit is a manifestation of the colonized personality in our politicians. All the popular administrations, such as *novo progresistas*, have been characterized by their contributions to this situation. The only party that has not yet had the chance to manage the Puerto Rican treasure is the *Independentista* party. We should ask ourselves what position they would take when dealing with the current structural deficit in Puerto Rico.

Economic self-sufficiency is possible outside the context of a colonized personality, something that colonized individuals fail to appreciate, given their dependency upon the metropolis and their fantasies of being rescued. An indication of a mature personality is keeping budget and expenditure in a balance based on available resources, thus avoiding the discomfort and anxiety that results from spending more than what is affordable. If electoral survival was not an issue, political parties would have already done away with the politics of increasing governmental payroll.

Our common goal should be to develop a thriving and robust economy, for which purpose we must invest money in the growth and expansion of our economy. Economic investment should be held in relation to work production. It is necessary to train and stimulate small businesses in Puerto Rico, since their development and expansion will in the long run generate more jobs and raise our national economic production.

The colonized government is jealous and envious of private entrepreneurship; hence, the two become rivals. This kind of government is characterized by excessively regulating private companies although in this colonial reality, local politicians do not really have much control over commerce. Commerce, rather, is controlled by the metropolis. The excessive internal regulation we find in Puerto Rico is none other than a mechanism that tries to compensate for a lack of real power to regulate free trade.

The United States imposes import and export tariffs between Puerto Rico and the United States, an obvious result of the fact that the United States possesses our territory. Economic self-sufficiency, on the contrary, should be founded on one's ability to conduct and control free trade that, through competition, results in lower prices.

The success of corporations such as Wal-Mart is founded on their ability to sell at lower prices. The reason they are able to do this is due to their ability to purchase products where production costs are lower. Thanks to globalization, many companies have survived and flourished. Similarly, free-trade opportunities within statehood, independence, or a nonterritorial free-associated state are sure to outnumber those found within our present situation.

"A full belly makes a happy heart." This is the slogan of colonized individuals. Their actions are always fully determined by present considerations, and, consequently, they become slaves to the present. The Puerto Rican economy has no growth plan and lacks any character of its own. The excessive governmental interference with the private sector tends to marginalize it, and our lack of economic planning hinders our view of the private sector as an indispensable factor in the development and growth of our economy.

We Puerto Ricans need to acquire a sense of direction that may lead us to the liberties we seek. We need to base our economic growth and development on free competition and the principle of merit. In order to develop a business, we must first become its owners. Once you own a business, the job acquires a new dimension as the feeling of owning the product of your labor reinforces the love and dedication you feel toward your work. But when the fruit of your labor falls into a stranger's hands, the love for the work and the pleasure gained by producing it is unavoidably lost. Colonized individuals have lost their love for work.

The large U.S. corporations have been promoted in Puerto Rico by large federal and state tax exemption incentives. As a consequence, many U.S. corporations have left their original states in order to relocate to Puerto Rico. This situation has been described by many governmental and private study groups as amounting to corporate subsidies. The exorbitant profits that they make, along with the fact that they had left their states of origin while scarcely benefiting the Puerto Rican economy with their investments, has resulted in the elimination of the laws that had originally engendered these corporations. This is a good example of bad economic planning by both the government of Puerto Rico and that of the United States.

People are our main resource in Puerto Rico; hence, our priority should be to intelligently take care of this human resource. By economically investing in that area, we will be able to further develop it toward a self-sufficient economy.

It is necessary to cultivate a corporate attitude among Puerto Ricans, which we can do through promoting incentives directed toward the growth of the largest possible number of small businesses. These can then become the core of our economy. Primordially, we must think about taking control of our lives and of our economy.

Education is the vehicle through which developing our human resources will become possible. Our education should be practical and relevant for solving life's problems. It should be focused on training students to solve problems. Reading, writing, and ably solving mathematical operations will prepare students for better jobs in this modern technological world.

Education should be directed toward contributing to the country's economy. Studying for the sake of studying does not make much sense. The value of education, rather, becomes apparent when students are capable of solving problems. Education that is focused on developing these three basic skills – reading, writing, and ably solving mathematical operations – makes sense.

Parents should require their children's teachers to teach the child how to read, write, and ably solve mathematical operations. With these basic skills, children will have a better chance of incorporating themselves within our present economic world. The Puerto Rican Secretary of Education is responsible for letting the public know which schools and which teachers are competent enough to teach those three basic skills. If this is not done periodically, then the only way to enforce it would be through public opinion.

In this global economy, thousands of jobs have been exported from the United States to other countries – where people can read, write, and ably solve mathematical problems – because of the lower costs. The competition of national economies nowadays occurs at an international level, and without the basic skills. As Puerto Ricans, we are unable to compete on this stage. Colonized people do not understand that their work and their education are the source of their self-esteem. We must become more responsible for our work and education in order to mentally decolonize ourselves. With the necessary rights and the fundamental skills, we should be able to magnify our worldview and direct our attention toward the self-sufficiency of our national economy.

Language is a dynamic thing – something we Puerto Ricans know well, given that most of us have been around Spanglish. Spanglish is one product of the dynamic and interactive character of language. Each day, languages and tongues are modified through human interaction, and this makes it possible for people who speak different languages to understand one another.

With an identity of our own, self-confidence, a desire to achieve economic self-sufficiency, and careful planning of our resources, the possibilities of progress for the Puerto Rican nation are endless.

In this global economy, having the United States as a business ally highlights the need to learn and use English as an essential part of our education.

CHAPTER FIVE

Predictions

As MENTIONED ABOVE, the great practicality of using the scientific method lies in the fact that it enables us to formulate hypotheses for predicting future events. The existence of a colonized personality among Puerto Ricans needs to be validated and confirmed by other behavioral scientists. This validation will be a long process that requires the effort of many scholars of human behavior. Many more observations by different analysts, and during the course of multiple occasions, will be necessary to validate this concept.

I will venture to make some predictions concerning the future behavior of Puerto Ricans, however, assuming the reality of the *colonized personality*. I will further assume that this personality is prevalent among a significant number of the population. My predictions will be based on those assumptions.

Prediction # 1
Puerto Ricans will continue to maintain the territorial *status* of Puerto Rico.

Prediction # 2
A consensus will not be reached among Puerto Ricans in demanding a change of the Puerto Rican territorial *status* from the United States.

Prediction # 3
The intervention of the United States Congress will be required in order to push Puerto Ricans toward political self-determination.

Prediction # 4
The colonized personality will reject *independence* as an alternative to the solution for the territorial *status* of Puerto Rico.

Prediction # 5
The colonized personality prefers *statehood* as the solution to the *status* of Puerto Rico.

Prediction # 6
The colonized personality will be the main obstacle for making Puerto Rico the fifty-first state of the United States of America.

Prediction # 7
For the United States government, the issue of the Puerto Rican *status* will eventually become an issue of civil rights.

Prediction # 8
The U.S. Democratic Party will take the lead in solving the Puerto Rican political *status*.

Prediction # 9
The process of decolonization of Puerto Rico will not lead us to civil war.

Prediction # 10
Throughout this process, the Puerto Rican nationality will reach higher levels of development.

References:

The original theme of *The World's Oldest Colony* comes from reading the book; The Trials of the Oldest Colony in the World by José Trias Monge, Yale University Press, September 1997.

End.

www.ingramcontent.com/pod-product-compliance
Lightning Source LLC
Chambersburg PA
CBHW021258280526
45784CB00005B/2424